THE HOUSEBUILDING HANDBOOK

Your pocket guide to building a low risk, high reward
property development business on a solid foundation

RICHARD & BRYNLEY LITTLE

RETHINK PRESS

First published in Great Britain 2019
by Rethink Press (www.rethinkpress.com)

Contents

Introduction

The Housebuilding Handbook is a practical guide taking you through the seven steps of the Housebuilding Business Blueprint™:

1. Sourcing

2. Appraisals

3. Securing the deal

4. Planning

5. Funding

6. The build

7. Exit

This handbook is primarily aimed at small and medium-sized enterprise (SME) developers who are looking to start, or scale, their business from a solid foundation of tried and tested methods, processes and strategies. It will focus on new-build property development, but much of the content is applicable to larger conversions. Whether you are looking to build a single house or large multi-unit projects, the *Housebuilding Handbook* will apply to both.

This book is for anyone who wants to create and develop a successful housebuilding business that will stand the test of time through ever-changing markets, have a positive impact beyond profit, and form the basis of a lasting legacy for generations to come, all while taking fewer risks. We will be giving our industry insights, revealing some of our methods, sharing stories and explaining through examples.

CHAPTER ONE

Solid Foundations

What is property development?

If you were to ask ten different people what property development is, you'd probably get ten different answers. Those who have been on *Homes Under The Hammer* and bought a dingy, rundown flat, giving it a new lease of life with fresh magnolia walls and an up-to-date kitchen and bathroom, likely consider themselves to be property developers. So would those who have converted a building into a house of multiple occupation (HMO) to rent out the rooms, and those embarking on bigger projects such as converting an office building into apartments or building a number of dwellings from the ground up.

If you've changed a property in some way and added value then it's fair to say you are a property developer, but it's clear that there are different levels, each with different complexity.

1. Refurbishment – taking an existing dwelling and improving it through structural and cosmetic building works

2. Redevelopment – extending, converting and/or changing the use of an existing building into single or multiple dwellings, obtaining the relevant consent in the process

3. Housebuilding – creating new dwellings from the ground up, building on previously developed, under-developed or undeveloped land

We consider property developers to be those creating new dwellings through conversion or new-build. Our focus from here onward is going to be on new-build property development, or housebuilding.

New-build property development is not 'another strategy' to those already in property investment, it is a different business entirely. Yes, there certainly are crossovers and skills transferable between the two, but it is not a business that all property entrepreneurs should get involved in.

Risk and reward

Management of risk is at the heart of any successful venture. This handbook goes some way towards helping you minimise risk. The most common types of risk associated with running a housebuilding business are:

1. Market risk
2. Financial risk

3. Technical risk

An example of market risk would be getting planning permission to build units that the market is already saturated with, such as blocks of apartments. Many inexperienced developers fall into this trap as an apartment scheme will produce the greatest number of units, which they associate with more profit.

An example of financial risk would be paying too much for the site in the first place, thus severely reducing your profit, or in some cases wiping it out entirely, causing big financial losses. Obviously you require significant sums of money to buy, build and exit new-build development projects. This can be a mix of your own funds, investor funds and lending from a development funder, so it's not just your finances at risk, yet you (as the lead developer) are ultimately responsible.

An example of technical risk would be spending time and money on the pursuit of planning permission with no planning strategy in place, leaving the end result to chance. You need to plan for each foreseeable outcome ahead of time to ensure you know what steps to take in any situation.

The single biggest risk to a house builder is getting involved in a bad deal. Generally, this comes as a result of a lack of understanding of:

- The market conditions and cycles
- The business principles and processes

Building houses requires a big investment, in both time and money, but to ensure you are going about things in the right way, the best investment you can make is educating yourself and surrounding yourself with experience, to minimise risk while maintaining and maximising your return.

Look after the risks, and the rewards will look after themselves.

Learning from experience

At what point would you consider someone to be 'experienced'? Would it be related to the number of deals they've done or are doing? How much profit they have made or are making? How long they've been in the industry? Or could it be related to the number of mistakes they have made?

In our opinion, an experienced developer is one who has been in business (risking their own money, not working for someone else) through at least one economic (boom/bust) cycle. If you are considering learning from, or working alongside, a developer to 'earn your stripes', align yourself with someone who has been through this journey. They will have robust systems and processes in place to ensure their business can survive difficult times and thrive in times of abundance. A softening market will expose developer mistakes that a buoyant market will cover up.

Boom and bust

The market is in a constant state of flux; some times are more volatile than others. One of the keys to a successful house-building business that stands the test of time is knowing when to act and when to sit back. It's about knowing when to focus on buying rather than building and holding rather than trading.

Generally, the best time to acquire land is when the market is soft, and the best time to be building out projects and selling units is when the market is on the rise. Inexperienced developers can often find themselves overexposed to changing market conditions. If you are building units at the wrong time, you could be looking to sell when the market has tipped. Market conditions need to dictate what types and sizes of projects you get involved in.

The conductor of the orchestra

As lead developer, you must become the organising force. The conductor of the orchestra is a great analogy to use, as like a housebuilding business, an orchestra has many different elements that must all come together at the right time. The conductor's job is to manage, oversee and organise the different elements to create a beautiful piece of music.

Your job as the lead developer is to bring together the necessary elements of the deal – the site, the money and the skills – and manage the process from sourcing through to exit.

More than money

Developers who are solely focused on making money are destined to fail. Property development is a slow-burn business, one that is capital intensive, with profit not materialising from projects for twelve to eighteen months or more. The most successful development companies have a clearly defined purpose above making money. For some, it is more about improving people's standard of living with quality homes. For others, it is creating a legacy through assets that they, their family and future generations can benefit from. One of the most valuable exercises you can do is to bring clarity to why you are building your property development business.

Profit will come as a result of concentrating on the following elements of the business:

- Purpose
- People
- Product
- Process

⚑ RED FLAG

At the end of each chapter, you will find a highlighted section called the **red flag**. This section **identifies** a common risk associated with that stage of the process, **understands** what implications it can have, and discusses a way to **mitigate** the issue to navigate you away from mistakes.

CHAPTER TWO

Source

I n this section, we will explore what opportunities to look for, how to find sites with development potential, the most common mistake inexperienced developers make when sourcing sites, and some top tips on how to approach landowners directly.

Sourcing sites is the lifeblood of your housebuilding business. It is a numbers game: you need to ensure enough opportunities are going in the top to produce the deals at the bottom. A common mistake inexperienced developers make is that they don't have a large enough amount of opportunities in their system to start with.

The panacea when it comes to sourcing is having multiple good prospects to process, allowing you prioritise them based on your available resources.

What to look for

Essentially, you are looking for anything where you feel value can be added. New-build development opportunities will fall into two main categories:

- Previously developed
- Yet to be developed

Many existing buildings occupy an underutilised plot. To open your mind when looking at opportunities with existing buildings on, ask yourself, 'Can it be knocked down?' This will get your creative mind looking at how to maximise the opportunity with a brand-new scheme.

In some instances, answering no to the question comes as quickly as asking it, for example a listed building, or a mid-terraced building that may have potential to build to the rear at a cost that would render many projects unviable.

New-build opportunities on previously developed land could include rundown houses, commercial buildings, agricultural buildings and blocks of garages.

Land that is yet to be developed could include large gardens (provided you can create a suitable access), agricultural land (depending on its location in relation to an existing settlement), and excess commercial land (such as part of a pub car park).

Where to look

There is a plethora of options available to you to source

opportunities. Below is a list of some of the avenues to look down to find sites:

- Rightmove/Zoopla/On The Market
- Estates Gazette
- Commercial agents
- Land agents
- Property/land sourcers
- Google Maps/Google Earth
- Paid mapping subscriptions (Nimbus/LandInsight)
- Property networks
- Business networks
- Family/friendship groups
- Construction/development professionals (architects/ project managers [PM]/quantity surveyors [QS]/etc)
- On the ground (walking/running/cycling your area)
- Local Planning Authority planning portal
- Brownfield register
- Social media
- Website traffic
- Paid adverts (digital/traditional)

You need to dedicate enough time to get momentum into your sourcing exploits. Focus the majority of your allotted sourcing time to the method(s) that will produce the greatest number of or best-quality leads.

A common mistake we witness with early-stage developers is a reluctance to do the sourcing themselves, opting to go through sourcers and agents. We understand that time is a precious commodity, and for some, sourcing isn't a natural fit, but we would urge every start up and early stage developer to source sites themselves until they've built up enough momentum and cash flow to increase their team, still keeping sourcing in-house.

With or without planning?

Adding value through planning is the biggest opportunity to maximise your margin. The best deals are those that either have no planning or the wrong planning in place. Although going through the planning process will extend the time of the project, the uplift in value usually more than makes up for it.

Have you ever seen development opportunities on the open market for twelve months or more with planning in place, yet they haven't been sold? There are countless 'deals' out there that aren't selling, which is often partly down to them having the wrong type of scheme consented. This offers you a great opportunity to connect with the owners and discuss looking at a scheme that would offer the greatest deliverable value.

The biggest mistake when sourcing sites

Many new developers, or those who haven't been in the business for a long period, focus on finding opportunities that are already on the market. They put a lot of time and effort into connecting with commercial agents, trawling the property portals looking for the next big deal.

There are three markets:

- On market
- The 'black book' market
- Off market

An opportunity will always start off market, with many landowners opting to connect with agents to market their site for sale. Every agent, whether land, commercial or estate, has a 'black book' of preferred clients and purchasers, and the vast majority of great development opportunities will be snapped up by these clients before reaching the open market. There are the odd exceptions to the rule, but your aim is to get yourself in the black book.

As soon as a site reaches the open market, every 'man and his dog' will become your competition. In any market where there is a lot of competition, the fear of missing out often takes effect and people get into bidding wars with other interested parties.

Dormant vs active prospects

Active prospects are those who are marketing their building or piece of land for sale. Dormant prospects are those who may be interested in selling now or at some point in the future, but not actively marketing. Although we don't ignore the active prospects entirely, our main focus within our own land-sourcing and development businesses is dormant prospects.

The number of dormant prospects far exceeds the number of active prospects. The best development deals are those that are off market, built on strong relationships with the key people involved.

The Sourcing Quadrants™

Every lead will fall into one of four categories:

- On market
- Off market
- Intermediary (a 'middle man' such as an agent or sourcer)
- Direct to owner (D2O)

Therefore, the four Sourcing Quadrants™ are:

1. Off market/direct to owner
2. On market/direct to owner
3. Off market/intermediary
4. On market/intermediary

	OFF MARKET	ON MARKET
DIRECT TO OWNER	1	2
INTERMEDIARY	3	4

No matter where your lead comes from within the Sourcing Quadrants™, your aim is to get it to quadrant one (off market/D2O). For example, if you find an opportunity on Rightmove, you are looking for an introduction directly with the owner through a series of conversations/viewings/meetings. In some cases, getting the agent to facilitate a meeting is the best way to do this.

—————————————— **CASE STUDY** ——————————————

We were interested in a small local development opportunity. Despite it being on market, we decided to send a letter to the owners to see if we could connect with them directly. We didn't want to depose the agent, or cut them out of the process; we just wanted to ensure we could talk directly to the decision makers to increase the chances of them accepting our offer. We suggested a meeting in the agent's office to discuss the numbers and our offer as a first step to moving the opportunity from on market to off market.

Direct to owner

These are the best deals you can do, so how do you connect directly with the owner(s)? In some cases, this can prove difficult with information not always being available, but generally speaking, the process of identifying, approaching and connecting with owner(s) is:

Step 1 – identify a site that has development potential.

Step 2 – locate the site on Land Registry with the owner's address or through the map search. (Get a Land Registry e-services account for free; it's a lot better than the standard Land Registry service.)

Step 3 – download the Title Register to reveal who owns the site and their registered address.

Step 4 – compile a letter introducing yourself, your expression of interest in the site, and a call to action to contact you. If you get no response within six weeks, follow this up with another (slightly different) letter. Repeat this with a third and final letter if there's still no response.

Step 5 – enter into a series of conversations/meetings to build the relationship, discuss the opportunity, dive into the owner's current situation, uncover any problems they are facing, and present solutions that you can offer.

Step 6 – once you have appraised the opportunity and worked out your numbers, put a proposal forward for the owner's consideration.

Step 7 – negotiate, ensuring that the deal works for everyone concerned.

Step 8 – close the deal and manage the relationship.

Written down, the process looks swift and simple, but in reality, deals can take a long time to agree. Make sure you invest enough time in building the relationship and cementing the connection you have, rather than trying to get the deal done as quickly as possible.

'Give me six hours to chop down a tree and I will spend the first four sharpening the axe.'

Abraham Lincoln

RED FLAG

Identify – becoming a motivated buyer.

Understand – being overly motivated to do the deal on any particular opportunity could lead you to making a decision fuelled with emotion rather than logic. This is more likely when you are at the early stage of your development career as you strive to get going or scale up.

Mitigate – the best way to prevent yourself from becoming motivated to buy is to build a pipeline of opportunities rather than focussing on a few. Having a pipeline in place allows you to emotionally detach yourself from specific sites and filter them properly through the appraisal process.

CHAPTER THREE

Appraise

The appraisal is the most important stage of the entire process. In this section, we will take you through the four levels of deal appraisal and discuss our 9 Pillar™ methodology, ensuring you are minimising your risk and maximising your return.

Get to the no quickly

No is the most powerful word when it comes to assessing development opportunities. Our appraisal process is designed to get to the no as quickly as possible, allowing you to focus your resources on profitable opportunities.

Our appraisal process has four levels:

1. Initial assessment
2. Planning assessment
3. High-level numbers
4. Full-development appraisal

These act as filters to sift out the right deals. Just like panning for gold, you've got to sort through the mud first.

Level 1 – initial assessment

When an opportunity lands on our desk, the first thing we do is conduct an initial assessment (IA) to establish whether there are any showstoppers that would bring an end to further assessment and allow us to move on to the next opportunity in our pipeline.

The 8-Point IA Checklist™ is:

1. Land Registry

2. Flood risk

3. Tree preservation orders (TPOs)

4. Access

5. Designation

6. Market history

7. Planning history

8. Local restrictions

The points don't have to be done in a particular order, and once you get into a rhythm of doing the IA, then you (or someone on your team) will be able to do it fairly quickly.

Land registry – check the title register see if there are any restrictions or covenants that may hinder or prevent development.

Flood risk – find out what flood zone the site is in (one to three) and check if it is at risk of flooding from surface water, rivers and seas, or reservoirs.

TPOs – these can severely affect the development potential of a site. If there are any trees, copses or hedgerows on or bounding the site, check to see if anything has a preservation order. Accessibility of this information will vary between different Local Planning Authorities (LPAs).

Access – this is the most common showstopper on the IA Checklist™. Whether you are looking to build one additional dwelling or larger multi-unit developments, you will need to establish whether the site has sufficient access. The speed limit, contours and designation of the road bounding your site will determine the dimensions of the visibility splay you need.

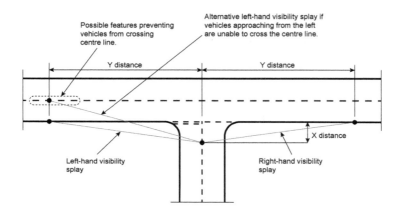

Possible features preventing vehicles from crossing centre line.

Alternative left-hand visibility splay if vehicles approaching from the left are unable to cross the centre line.

Y distance

Y distance

X distance

Left-hand visibility splay

Right-hand visibility splay

CASE STUDY

A local contact brought a site to us. He was interested in refurbishing the existing property and building a new dwelling in the rear garden. This property was on the market for sale and our contact had already had his offer accepted.

The refurbishment of the existing dwelling on its own would not have been profitable; the 'gold' was in the garden. He wanted our thoughts prior to exchanging contracts, but unfortunately, we didn't give him the news he was hoping for. On the face of it, there was enough road frontage and space within the site to create access for the new dwelling, but we uncovered a big issue when looking at the access in more detail.

Because an additional dwelling would increase the traffic movements, we had to look at the junction serving the existing properties. The speed limit of the road was 30mph, meaning we had to have a clear visibility splay 2.4 metres back from the road edge of 42 metres in both directions. To the left, the visibility splay crossed over someone else's land, and to the right it had a mains electric transformer slap bang in the middle. Although it would have been possible to have the transformer moved, the substantial cost would have made the new-build project commercially unviable.

Thankfully our contact didn't exchange on the deal.

Designation – this refers to the designation of the area and/or building, such as greenfield, greenbelt, brownfield, employment zones, local centres, conservation areas, retail, etc. The area designation can in some cases restrict or prevent new development or change of use.

Market history – check to see if the site is, or has been, marketed for sale. This isn't a showstopper but gives you a potential indication as to the motivation level and expectation of the seller.

Planning history – go on to your LPA planning portal and see whether the site has any current or previous planning history. Has an application ever been submitted, withdrawn, granted, or refused?

Local restrictions – look for physical things in the locality that may affect future development, such as overhead cables, bus stops, phone boxes, double yellow or red lines, and overlooking windows.

If you fail to find any showstoppers, continue your appraisal on to level 2.

Level 2 – planning assessment

A planning assessment is a deeper level of analysis concentrating on the planning potential and constraints. At this stage, you are looking at what may be possible in terms of a new scheme for the site and what is likely to restrict the type, size, number and layout of any new dwellings. Orientation and proximity of windows on neighbouring properties is a great example of a constraint that could affect development.

What surrounds your site can play a big part in what you can build. This can work to your advantage if you are looking at a site with nothing to the side. Clever scheme design can restrict what is built next to your site and protect, and potentially enhance, the value.

A typical planning appraisal put together by a professional includes the planning history of the site, planning history of any other relevant sites, what planning constraints there are, a review of both national and local planning policies that will apply to any potential scheme, and extracts from the local plan and supporting documents. At a glance, you are looking to find information to give you an indication of what types of units you can build and how many. Armed with this information, you will then start looking at the numbers to aid early conversations with agents and land/site owners.

Level 3 – high-level numbers

High-level numbers give you an idea of ballpark values and costings prior to doing detailed breakdowns and cash flows. This level of analysis can be helpful in early conversations with landowners and agents to gauge their expectation verses reality.

One of the biggest hurdles you will have to overcome is owner expectation of what their site is actually worth. Your ability to manage expectation and handle objections will determine how successful you will be.

Working on high-level numbers involves a simple spreadsheet covering the main areas of costings associated a development project:

- Gross development value (GDV)
- Build cost
- Contributions
- Professional fees
- Finance
- Developer's profit
- Residual site value

GDV – once you have an idea of how many units you are likely to build, work out an achievable £/m² sales value by researching comparable sold prices in the area, dividing the end sales value by the square meterage of the property. Multiply the £/m² value by the total m² you are intending to build to produce a high level GDV. Work in square feet or square metres, whichever is better for you. 1m² = 10.76ft²

Build cost – assessing build costs is one of the most common areas of uncertainty for early stage developers. It's dangerous to base offers on rules of thumb, but you have to start somewhere. Build costs can range from sub £800/m² to in excess of £2,000/m². For a 'standard' new-build development scheme, we often start with a figure of £1,200/m² as our ballpark build cost, adding a 5% contingency plus specific amounts for any abnormals and utility provision.

Contributions – depending on the type of development, the

LPA may collect a contribution from you as the developer. The most common are section 106 (s106) and community infrastructure levy (CIL). Affordable housing falls under s106 and your LPA will specify what percentage of affordable units will be required and/or what sum of money you will have to pay as a contribution. CIL is a levy applied to newly built space charged per m^2, published on your LPA website.

Professional fees – a raft of professional services will be required on all new-build development projects, such as architect, planning consultant, structural engineer, QS and PM. Fees typically range between 8–12% of the build cost.

Finance – the main elements of funding are the land purchase, the build and the holding costs. Where to go to assess these and how much you'll pay will be dependent on multiple factors, including market conditions, the nature of the scheme, your experience and the team you have around you.

At this early stage, keep the funding element simple and allow for a 10–12% interest rate with a straight-line cash flow. The resulting figures will not be 100% accurate but are good enough to work out a headline cost. We will consider more detailed cash-flow models later in the process.

Developer's profit – profit is measured in different ways, such as return on capital employed (ROCE), return on GDV

(ROGDV), internal rate of return (IRR). Our target profit margin is 20% ROGDV.

Site value – what is left after you've considered all of the above is the residual value of the site.

--------------------- **CASE STUDY** ---------------------

Here are the figures, rounded up or down for ease of explanation, for a nine-unit residential new build.

GDV – 800m² (total build area) × £2,800 (per m² sales values) = £2,240,000

Build cost – £1,200/m² + 5% contingency + utility provision = £1,035,000

Contributions – CIL £58 per m² + s106 contribution = £60,000

Professional fees – 10.5% of build cost = £108,675

Finance (build) – 10% pa interest for 7.5 months = £64,690

Developer's profit – 20% ROGDV = £448,000

Gross site value (exc. finance for purchase) = £523,635

Finance (purchase) – 10% pa interest for 15 months = £65,455

Residual site value = £458,180

The purpose of this level of number crunching is to allow you to filter through opportunities, gauge expectation and prioritise leads that you feel have sufficient margins, not to make firm offers. As you become more experienced you will be able to start looking at sites very quickly and apply high-level numbers based on guided percentages.

Level 4 – full-development appraisal

We've now reached the business end of the appraisal process. A full-development appraisal, as the name suggests, is an in-depth analysis of the development opportunities that have filtered through the early stages of the appraisal process. We look into nine different areas to establish whether to move forward with a formal offer/proposal. We refer to the nine areas as the 9 Pillars™ as they provide the strength to our appraisal.

The 9 Pillars™ are:

1. Demand

2. Scheme design

3. Planning

4. Legals

5. Construction

6. Funding

7. Costs

8. Deal structure

9. Exit

Demand – what demand is there for certain unit types in the area? We see all too often inexperienced developers opting for blocks of apartments in areas where there is already an oversupply of that unit type. Just because one particular scheme creates the largest amount of units or highest GDV doesn't mean it is the optimum scheme for the site. A good quality estate agent is worth their weight in gold; engage them early in the process and get the inside track on what units are selling and renting well, and their thoughts on realistic sales values and timescales.

Scheme design – it's important to compare different schemes and run them through your appraisal spreadsheet to decipher which is the best one to take forward. There's no need to invest £1,000s on expensive drawings from an architect at this stage; in fact, it is better initially for you as the lead developer to produce basic scheme layouts for your architect to improve upon. Block plans, m² or ft² based on national space standards, and density calculations are sufficient to assess what types of units and how many are suitable for your site.

Planning – any risks surrounding successfully obtaining planning permission will be mitigated in the structure of the deal. The focus on planning, aside from getting an idea of what you could achieve, is about understanding the potential timeframes involved and the strategy to achieve a successful outcome. A planning consultant is an indispensable part of your team and early dialogue will make for a smoother process moving forwards.

Legals – legal advice prior to purchase will establish any technical risk associated with the deal that may affect costs and timeframes. Legals weave in and out of the entire appraisal, from deal structure to exit.

You will often need title reviews at this stage to assess the implications if there are any easements or restrictive covenants on the title. It's better to invest a smaller amount of money upfront than uncover issues further down the line.

Construction – how you are going to procure the build will have an effect on the costs, funding options and your potential margin. For many, employing a main contractor is the best option, especially during the early stages of a new build career. At this stage, you don't need to know the finer details of how the site is going to be built, but consider the construction method as this will have an impact on costs and timeframes.

Funding – access to funding hinges on the strength of your appraisal. You need to ensure you have satisfied your lender in order to receive the necessary funds to proceed with the project.

Not satisfying funders is one of the biggest reasons deals fall out of bed, which is a great reason to keep track of 'sold' deals as they may become available again. When you're working out your funding costs, apply achievable rates and timeframes. Projects running over time can be incredibly costly, which erodes your profit. As well as your interest rate, you need to consider entry, exit, valuation, broker and cost-management fees.

To obtain a more accurate cost of borrowing, have an idea of your cash flow: when and how much money is coming in and going out. Apply a 'burn rate' (the rate in which you are spending) to your numbers to see your negative cash flow.

For example, if you purchase a piece of land for £250,000 for a project lasting twelve months, you would require £250,000 for the full duration, so the burn rate is 100% (all the money for all the time). At an interest rate of 10% pa, the cost of finance would be £25,000.

Estimations on sales (when and how much) show your positive cash flow. Early (off-plan) sales, especially on bigger projects, are paramount to reducing your borrowing costs and increasing your margin.

Costs – for those of you who love spreadsheets, this is the time to get excited. Numbers obviously play a big part in what makes a good deal. Where a lot of developers slip up is having a spreadsheet with only one column of numbers. How many costs involved in the development process are fixed costs? The vast majority of costs associated with development will vary. Our spreadsheet has multiple columns designed to create a range of numbers.

The purpose of creating a range of numbers is to establish where the biggest risks lie in terms of accurate costings. You will have a low-range column, a high-range column, a proposed cost (most likely) and actual cost (to be filled in after completion of the project).

Low	High	Proposed	Actual

As you go through each line item and assess the costs, investigate any that have a big differential between low and high prior to completing on the deal.

————————————————— **CASE STUDY** ————————————

We were working on a site for seven new builds. There was an existing gravel track road leading from the main road to where the units were to be located. One of the big unknowns was how much the road was going to cost to upgrade. Without having extensive ground investigations done, we had to assess the likely costs of two scenarios:

1. Replacing the top coat of the existing road, assuming the bottom surface was sufficient and up to current standards

2. Building a completely new road, including excavation of the existing road

Our estimations varied from £40,000 to £150,000, giving us our low-range and high-range figures. The sheer difference between low and high represented financial risk and became a condition of our eventual offer, subject to ground investigation to allow us to determine a more accurate figure before committing to a specific number. If the landowner hadn't been prepared to share the risk, then we'd have taken the higher number when working out our eventual site value.

Costing a project is a big cause for concern among many early stage developers. Input from QSs, contractors and development consultants will give you the confidence to move forward knowing you've not made any grave mistakes on the numbers.

Deal structure – the structure must be fair and reasonable for all parties involved. You need to establish how you propose to purchase the land: whether you will be using a control contract such as an option agreement, or whether you will be looking for a conditional or unconditional sales contract. The structure of the deal will affect your funding options, your costs and the amount the landowner could receive.

Inexperienced developers often want to enter into joint ventures with landowners and investors. In our experience this rarely works out well and has knock-on effects during legal reviews and funding.

Exit – it is important to have a clear understanding of your exit plan during the appraisal stage to de-risk as much as possible. Ultimately your deal is only successful once you have exited, whether that be achieving sales, refinancing to retain units yourself or with partners, or a combination of both. You want to establish where your tipping point is: the point at which you are out of debt. High-value schemes, for example four £800,000 executive homes, carry a huge amount of risk as you often won't reach the tipping point until every unit has been sold. This puts a large amount of pressure on early sales.

If you think a good deal is determined by numbers on a spreadsheet, then you've missed the point. A good deal is about creating value and finding a balance between what is permissible (by the LPA), acceptable (from the perspective of the landowner and funders) and desirable (for you as the developer and end users).

⚑ RED FLAG

Identify – not spending enough time, money and effort appraising deals.

Understand – new and early stage developers often want to rush the appraisal process and get offers in as quickly as possible. This can be fuelled by the fact that they are looking exclusively at on-market deals and competing with others. Rushed appraisals lead to superficial data, which results in unqualified offers. You never know, some may come off, but in most cases, you pay too much for the site and enter a world of problems.

Mitigate – spend more money at the front end (sourcing, appraising, securing) to save you time, money and stress further down the line. Ensure you have a sufficient pot of working capital to begin with for pre-deal costs such as marketing, lead generation, expert guidance and advice from your LPA (if available).

CHAPTER FOUR

Secure

The next stage is to secure the deal. This will involve putting together proposals, discussions and negotiation. In this chapter, we will be explaining the most common contracts and agreements used in land and development purchases and discussing negotiation best practices.

One of the core values that runs through all of our businesses is *fairness*, so we like to ensure any deal we do delivers value and creates a win-win. This is a quality that will stand you in good stead when building a successful, sustainable development business. Creating deals that work on the selling side of the transaction comes down to your ability to understand the owner's present situation: what they'd like to achieve, their motivations and any inhibitors.

Agreements and contracts

You can be creative when it comes to deal structures, but you need to ensure that you don't oversimplify or overcomplicate matters when it comes to contracts and agreements.

Essentially there are two types of contracts used in acquiring development opportunities:

- Control contracts
- Purchase contracts

Control contracts give you as the developer rights over land you do not (yet) own, allowing you the opportunity to invest in adding value, usually through planning prior to purchase or sale. The most common control contracts are *option agreements* and *promotion agreements*.

Option agreements have been around for centuries and are often the agreement of choice for both small and large developers for sites without planning in place. They allow you the option to purchase without obligation, mitigating the risk of having to purchase a site regardless of whether you obtain planning consent successfully or not.

Promotion agreements are commonly used on sites without planning permission. With this type of agreement, the developer (or land promoter) is prepared to take the financial risk of obtaining planning permission, and the landowner is prepared to share a portion of the resulting uplift in value with them. Land promotion usually takes place on strategic pieces of land on the edge of urban and rural settlements.

Below is an example deal done under a promotion agreement.

EXAMPLE DEAL

A developer identified a site using Google Maps and contacted the owner through direct mail. They set up meetings, negotiated terms and signed an agreement. We have rounded the numbers up or down for ease of explanation.

Total site area = 2.15 acres

Existing use value (pre-planning) = £25,000

Cost of planning = £40,000

Enhanced land value = £575,000

Net uplift = £510,000

Landowner share (75%) = £382,500

Promoter share (25%) = £127,500

Control contacts often lead straight into purchases once you've created an uplift in value. Purchase contracts are typically used when it's easier to establish the set value for the site.

Purchase contracts fall into two categories:

- Unconditional purchases
- Conditional purchases

Unconditional purchases are straightforward purchase agreements at a set price. They are rare in the development world, although requests for unconditional offers are not uncommon when deals are on the market.

Conditional purchases are much more common as they account for the many moving parts that make up the deal. Conditions to an offer could be:

- Subject to planning permission
- Subject to ground investigation
- Subject to developer contribution changes
- Subject to access

The conditions usually reflect the risk associated with the deal.

An overage clause often gets added to land purchases where the landowner and developer plan to share any increase in value over a set amount. For example, if the deal was to purchase a piece of land for £250,000 based on a GDV of £1,000,000, but in fact a GDV of £1,200,000 was achieved. The addition of an overage clause would mean the developer and landowner would share the additional £200,000 in revenue.

Negotiating the deal

We often hear the words 'I can't get the deal to stack up' from developers we are connected with. What they tend to mean is 'The owner's expectation is too high, and I can't negotiate them down to a realistic value.'

The art of negotiating a development deal is about truly understanding what is motivating the owner to consider selling their asset and bringing clarity to the conversation rather than confusion. Two of our best practices when it comes to proposing, negotiating and closing deals are:

- Multiple offers
- Open book offers

Multiple offers – tabling more than one offer increases your chances of one of them being accepted. You need to make sure all your offers work for you, and that they have different benefits for the owner.

---------------------- **CASE STUDY** ----------------------

We had the opportunity to purchase part of a pub car park to build four two-bedroomed terraced houses (great rentals). By looking back through the previous planning history of the site and talking directly with the owner, we soon established their motivation for selling was because they required funds to pay for necessary works on the pub to keep it open. The amount they needed was £50,000, so we structured two offers that fulfilled their needs. Offer one was a straightforward purchase at £110,000, and offer two was £50,000 (the amount they required) with a further payment of £90,000 on completion of the development, taking the total purchase price to £140,000. They accepted offer two.

Open book offers – involve you discussing your breakdown of numbers with the owner and/or agent. This raises your credibility and reinforces your level of expertise. By going through the main sections (as per your high-level numbers), you can discuss specific areas and focus on the areas of disagreement rather than have them dismiss the offer in one go. This adds value to the owner/agent and shows them that you are professional.

☐ RED FLAG

Identify – being too creative with your deal structure.

Understand – just because you can doesn't mean you should. A good example of being creative without understanding the implications is something we see many inexperienced developers attempt:

entering into a joint venture where the landowner 'puts the land in', the developer funds the build, and they share the profits. While on the face of it, these types of deals make sense, they rarely come to fruition once you reach the legal phase, and often seriously hinder your ability to borrow money from development lenders.

Mitigate – keep the deal structure as simple as possible and ensure that how you are looking to proceed will be acceptable from a lender's perspective.

CHAPTER FIVE

Planning

Many view the planning system as red tape and something they have to fight their way through. These people don't last long as developers.

This section is not a deep dive into the intricacies of the UK planning system, but rather an insight into understanding how to work with the system and use it to your advantage. We are going to give you a brief overview of the planning system in England, looking at the key people involved throughout the process and focusing on planning strategy.

Please note that if you are developing in Wales and Scotland, there are differences in the systems of those countries compared to that of England.

The planning system

The National Planning Policy Framework (NPPF) is the overarching document that contains the guidelines for all LPAs to set local policies and control development. The NPPF, set out by the national government, contains topics such as achieving sustainable development, delivering a sufficient supply of homes, building a strong, competitive economy, making effective use of land, and achieving well-designed places.

England has a plan-led system, meaning every LPA must prepare a local plan containing local policies and an outline of how land should be used, determining what should be built when. Local policies are typically set out over a ten- to twenty-year period to ensure development is in line with creating sustainable infrastructure and communities. By law, each local plan must contain core policies, site allocations development plan document and a proposals map. In addition to the local plan, LPAs produce supporting documents such as strategic housing land availability assessments (SHLAA) and strategic housing market assessments (SHMA) that focus on the suitability, availability, achievability and need of land for housing and development.

You may find that your LPA has supplementary documentation relating to design guidance. These are useful when it comes to looking at scheme design and how the land can be used most efficiently. You'll find information identifying criteria that local projects should adhere to such as car parking, amenities, open space, local character, and density.

Developer contributions

Financial contributions are collected by the LPA most commonly in the form of s106 and CIL. These contributions are designed to mitigate the impact of new homes (and other buildings) which create extra pressure on existing infrastructure and local facilities.

S106 – the three purposes of this planning obligation are:

1. To prescribe the nature of the development
2. To compensate for loss or damage created by a development
3. To mitigate a development's impact

One of the most common, and most impactful, obligations under s106 is the provision of affordable housing. Affordable housing contributions are provided in one of two ways:

- On-site provision
- Financial contribution

LPAs will stipulate the percentage of affordable homes required to be provided on developments of certain sizes. Parameters vary from council to council, so it's important to be clear on what the threshold and percentage are in your area.

For example, if you are planning to build a development of twenty dwellings and the LPA stipulates an affordable housing provision of 40% for developments over nine dwellings, then eight of the dwellings will need to be provided under a prescribed mix of affordable for sale and affordable for rent.

Affordable housing is a complex part of development. If you want to operate at scale, understanding its intricacies, harnessing its potential and embracing the possibilities will be pivotal to the growth and impact of your business.

Obligations under s106 can also include contributions for education, trees, ecology, recreation, open space, public art, and transport. S106 agreements often require you to make financial contribution prior to commencement of works. If this is the case with your projects, make sure you allow for this when working out your cash flow.

CIL – a tariff-based planning charge brought in as a tool for LPAs across England and Wales to help deliver infrastructure to support growth and development of local areas. Your development may be liable for CIL if your LPA has chosen to implement a charge in its area. The levy may be payable on developments which create net additional floor space of 100 square metres or more. That limit does not apply to newly built houses and flats, and a charge can be levied on a single dwelling of any size. Exclusions, exemptions and reliefs may be available.

Levy rates are expressed as £ per square metre. A mistake developers often make when calculating the financial impact of CIL is not accounting for the indexation charge set out in the charging schedule. The original published rates increase each year based on a percentage set out in the schedule, so make sure you've accounted for the increase in your numbers.

Planning strategy

Planning is like playing a game of chess. In chess, you make moves to achieve the desired objective of 'check mating' your opponent's King. In planning, you make a series of moves with the objective of achieving consent for a development you (or another developer) are looking to build. Good chess players who anticipate their opponent's moves are usually three or four moves ahead of the game and play strategically while also having the ability to react to situations to come out victorious. Approach planning like a game of chess. There are lots of different ways to reach the winning post, so first determine your strategy to get there.

To come up with an effective planning strategy, you need to bring together technical knowledge and the ability to influence and manage the people involved in the process.

The process – in order to be prepared for all eventualities, it is essential, as a lead developer, to have an understanding of the planning process from start to finish.

Below is a high-level seven-step planning process:

1. **Advice** (pre-application)

2. **Application** (outline, planning in principle [PIP] or full)

3. **Acknowledgement** (validate or request more information)

4. **Announcement and assessment** (publicise plans and consultation period)

5. **Answer** (decision from planning officer or planning committee – granted, granted with conditions, refused, non-determination)

6. **Appeal** (right to appeal the decision)

7. **Ask** (Judicial review – right to challenge a decision in the High Court)

The key players – understanding who is involved throughout the process and the role and responsibilities they have is crucial to coming up with a planning strategy that will bring you the result you are looking for. Members of your core team (internal team and external consultants) need a good understanding of other recent planning applications, decisions and appeals. In-house planning experience/expertise is obviously a plus, but planning is an area where the majority of input is brought in from outside.

Your planning team would generally be made up of:

- Yourself (as the lead)
- Planning consultant
- Architect (for the design element)
- Specialist consultants (such as highways, environmental and civil engineers)

On the other side of the fence, the key people/groups involved are:

- Planning officers (planning policy and development control)
- Planning committee
- Parish/ward councils
- Local residents
- Third-party consultants
- Planning Inspectorate (appeals)
- High court

To give yourself the best chance of reaching a satisfactory outcome, adopt certain best practices to manage and influence the 'opposition'. Ensure you maintain communication with planning officers and councillors. Relationships are the key to everything in this business. Profile the members on the planning committee to give yourself (and your team) deep insight into voting patterns, contributors, and what schemes typically receive support or resistance.

The strategy – as the saying goes, 'There's more than one way to skin a cat', and with planning there are lots of different approaches to achieving the desired result. One example of a strategic approach to planning would be having multiple applications going through the system for the same site. You could opt to run applications in parallel, phased or end to end. We have done this previously, the first application helping us to understand what the objections are, the second application meeting the requirements. As in chess, sometimes the right move is to sacrifice your pieces in order to gain a position of strength.

The important thing to note is that policies, personnel procedures and plans are always under review, causing the system to be in a constant state of flux. This reinforces your need to have an experienced team of people around you who have their fingers on the pulse and are strategic, methodical and proactive in their approach to planning.

⚑ RED FLAG

Identify – having unrealistic timeframes for the planning process.

Understand – getting your timescales wrong can have huge knock-on implications for the rest of the process. We have seen developers run over on their agreed finance terms due to planning, resulting in expensive penalties.

Mitigate – discuss realistic timescales with your team early on in the appraisal process. Timeline all possible outcomes in relation to your application and have plans in place if the decision goes to committee, to appeal or is refused.

CHAPTER SIX

Funding

I n this section, we'll go through the funding options and structure, cash flow, fundamentals of raising finance, and take a deeper look into the risk money mindset.

The need for funding runs throughout the entire process of property development, from your early stage business processes such as lead generation, marketing, and general business overheads, to site acquisition, planning, the build phase and exit funding. To run a sustainable, scalable business, you first need good working capital to get you to the point of site acquisition, which is where more opportunities for finance open up.

Funding options

There are tons of options out there to fund your projects, but lack of experience can cut down the options available and affect the rates you are likely to pay. Below is a list of funding options for acquisition, planning and construction:

- Your own money
- Investor funds
- Equity (in businesses, property or other assets)
- Banks
- Development finance companies
- Bridging loans
- Peer to peer lenders

There are lots of iterations within these, but the list makes up your main options.

Financing a project can take many forms with funding solutions tending to be deal specific, considering different factors. A typical breakdown to fund a development project could be:

Land – purchased outright using your own funds, investor funds (on a fixed rate loan basis), or a combination of both.

Development – the build would then be funded using a loan from a specialist development finance company. As you have purchased the site outright, you are likely to be able to borrow 100% of the cost of the build. If 100% isn't available, then you will have to plug the gap.

Exit – sell the units on, allowing for a sufficient holding period while sales are completed.

Because projects can be funded in various different ways, it's important to understand the funding structure or 'capital stack'.

The capital stack

There are three parts that make up the capital stack.

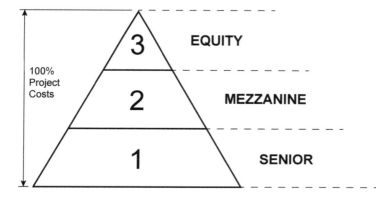

1. **Senior debt** – typically a specialist development finance or institutional lender who takes first charge (the clue's in the name). They usually fund up to 65% of GDV or 80% of total cost, up to 70% of the purchase price and up to 100% of the cost of development. The cost of capital ranges from 5–15% per annum.

2. **Mezzanine** – comes from a specialist lender or private funding, holding a second charge over the asset. 'Mezz'

funding plugs the gap (if there is one) between the senior debt and equity up to 90% of the total project cost. This type of finance is more expensive than senior debt with cost of capital ranging from 15–25% per annum.

3. Equity – this is your 'skin in the game' coming from your own funds, or investor funds on a fixed return or profit share. The equity funds the balance of the capital requirement under third or no charge.

The makeup of your capital stack will depend on factors such as your own available funds, your experience level, your investor pipeline and the quality of your deal.

Raising finance

Whether you are looking to raise finance from private investors, companies or both, there are some fundamentals to go by to ensure you are attracting money rather than asking for it. We call it the MONEY MAGNET™.

Mindset – money is nothing more than a commodity you need to bring a project to fruition. The same can be said about bricks. You need many £s to fund a build and many bricks to construct it, yet developers tend to focus on shopping around for the best prices for bricks rather than money.

Negotiating rates on your finance raises your credibility, even more so when you are bringing a good deal to the table packaged with a good project team. A quarter of a per cent off your finance can equate to a substantial amount that goes straight on your bottom line.

Add value – if you want to attract investment consistently, add value to others in the form of useful insights, educating your audience. Rather than going cap in hand asking for investment for your projects, raise awareness, building your authority and proving you know what you are talking about.

Green light – are you in a position to move forward if you're given the green light? Make sure you have the required documentation ready to send to your prospective lenders and investors, for example your appraisal, team details, track record, site details/plans, marketing plan, etc. Getting into a habit of setting up folders containing all the necessary documents and being able to provide them quickly shows that you are organised and structured. Money likes structure.

Network – your personal and professional networks are key to getting funding both directly from investors and through introductions to finance companies. The best investor relationships are ones that are built over time. A big mistake many early stage developers make is concentrating their efforts purely on property and business networks, neglecting those who already know, like and trust them.

Ethics – operating your business under a strong code of ethics will take you further than anything. It starts with getting clarity on why you do what you do and what's important to you, followed by amplifying that through your thoughts, words and actions. When you are seen to be working towards something greater than yourself and your own personal goals, making a difference, people will be attracted to be part of that. Become purpose and value driven.

Track record – when you're starting out without the benefit of a track record of completed projects behind you, your focus needs to be on assembling a project team with the relevant experience to gain funding and instil confidence in your lenders. This team is likely to play a big part in your early stage career, and who knows? They may even stay with you throughout your ascent to experienced status.

As you build up experience, documenting it is key. You can then use facts, figures and photos to demonstrate your ability to deliver to potential investors.

However you are raising money, you need to make sure everything is legal and above board, and you are complying with the relevant rules and regulations.

Managing your cash flow

Being able to manage cash flow is a success-defining responsibility in most businesses, but especially in housebuilding as it is so capital intensive. A lot of capital is required to bring a project through the whole process, so it's important you, as the lead developer, have a clear understanding of the money flow in and out of the business.

Typically, your cash-flow curve will be steep at first due to heavy early stage costs such as site acquisition, planning and groundworks. It then flattens out during the main build phase with outgoing payments becoming more regular. Finally, you'll see an increase in spending as the project gets closer to the end with expensive second-fix purchases like kitchens and bathrooms.

Cash coming into the business as early as possible really helps your cash flow. This is why developers of multiple units look to secure off-plan sales. Bringing cash in early is a big win for developers, which is why you *can* afford to offer discounts for early sales.

! RED FLAG

Identify – not having a 'float'.

Understand – because development finance is paid in arrears and not in advance, if you haven't accounted for a pot of money to kick the project off, then you either won't get the offer of finance from the lender or will be chasing your tail from day one, scraping together money for professional fees and initial works by the contractor.

Mitigate – when assessing how much equity you will need to inject into the project, always allow for a float to cover initial outlays, some of which will be recovered from the money you draw down from your lender.

CHAPTER SEVEN

Build

For many the construction phase is the most exciting part of the process: seeing dreams become reality.

During this section, we will be running through the construction phases, procurement and management of the process, and the benefits of new-build development over conversions.

There are three phases to the construction process:

1. Pre-construction

2. Construction

3. Post-construction

Pre-construction

The construction process starts way before you have workers on site and break ground for the first time. Pre-construction is arguably the most important construction phase as the better prepared you are, the more chance you have of bringing the project in on time and on budget.

Activities in pre-construction include design team management, utility management, consents and permissions, scope of works and pricing, tendering and contractor selection, pre-contract administration and programming. Pre-construction is usually a role taken on by a PM.

Utility management can have a domino effect on your timeframes. By not engaging with utility providers early enough in the process, you may have to put other tasks on hold, causing early delays. Utilities can also hit your pocket hard if you haven't allowed enough money for them in your financial appraisal.

To mitigate as much risk as possible, find out the providers in the area and engage with them early in the process to obtain information and budgets.

Construction

The construction phase accounts for the majority of expenditure in terms of both time and money, so procurement and management are two vitally important operations that will protect your profit and maximise your margin.

There are many different routes by which you can bring about the design and construction elements, and your procurement strategy should fit your long-term objectives, taking into consideration cost, speed, quality, risk, experience and financing. The 'traditional' procurement method is where the client appoints consultants to design the project in detail and then prepare tender packs to invite contractors to submit tenders on a competitive basis. This method of procurement is suitable for both experienced and inexperienced developers.

If you are looking for a more hands-on approach, you could consider becoming the main contractor yourself, dealing with multiple trades to deliver specific parts of the build. The benefit of becoming the main contractor is that you can increase your margin further. As a general rule, main contractor pricing comes in with an approximate 20% uplift, but that 20% comes at a cost. Increased reward generally means an increase in resource and risk. There is an approximate cost of 8–12% when you become the main contractor, which results in a saving (or gain, depending on how you look at it) of 8–12%.

Please note that becoming the main contractor is not for everyone, especially if you're in the early stages of your development career. However, for those with existing industry experience and/or looking to scale their business, it could be the best route to go down.

Post-construction

Reaching the end of a project is always exciting, but there are still things to take care of. Post-project completion, you will need to set a period of time to allow for any defects to be fixed by the contractor. Some defects, although not visible at competition, may appear during use of a building and can be rectified, thus releasing any retention fund.

Defects are a common cause of disputes on projects as it is not always clear what constitutes a defect and who is ultimately responsible. It is generally the job of your PM to handle post-construction tasks, including certification with building control and warranty providers, management of the agreed defects period, disputes, release of retention funds and sign off of the final financial account. If there are elements of the scheme that do not fall under the responsibility of the owner and occupiers, such as communal areas in a block of apartments, or un-adopted roads and open spaces on private developments, and are not brought under the control of the council, then you will need to consider how you are going to hold and manage or whether you can sell to specialist companies.

New build vs conversion

One of the main advantages new builds have over conversions is cost control. The difficulty with costing conversion projects is that there are often unknown costs throughout.

When it comes to costing a new-build project, the big unknown costs are in the ground (foundations, services, etc). Once you are out of the ground, you can attain accurate build costs with a small allowable contingency (which is not there to be spent). One of the most common reasons for overspend and the depletion of your contingency is specification changes.

In terms of risk, with new-build schemes of houses, as opposed to large-scale conversions and new-build apartments, it's easier to phase and control the delivery of units when you're dealing with changing market conditions.

⬜ RED FLAG ─────────────────────────

Identify – going over time and over budget, ultimately hitting your pocket.

Understand – delays in the build can result in bringing properties to market at the wrong time, losing sales, and increasing your finance costs. Every £1 over budget is effectively £1 off your margin.

Mitigate – assess realistic timeframes and budgets during the appraisal stage and set them during pre-construction. As early as possible, highlight things that could impact the project length and cost, for example site surveys, discharging planning conditions, and utility provision. Employ the services of a good PM who's experienced in pre-construction as well as on-site management, and get your project team together regularly to ensure everything is on track. Management of the process is key to maintaining margin.

CHAPTER EIGHT

Exit

An understanding of your exit – what you are ultimately looking to achieve – will impact how you go about the preceding process, so it's imperative to get clarity prior to acquiring sites. The ideal scenario is to have multiple exit routes available for deals that pass through your pipeline.

The four exit strategies are:

1. Deal packaging

2. Planning uplift

3. Build to sell

4. Build to rent

'Begin with the end in mind.'

– Stephen Covey, *The 7 Habits of Highly Effective People*

Deal packaging

Sourcing projects and selling them on to other developers is a great way of bringing cash into a capital-intensive business. Once you have a steady flow of leads into your pipeline and a process set up to filter the wheat from the chaff, you will need to assess your capacity in terms of the number of projects you have the resource to work on effectively yourself.

Let's say your capacity is three projects a year. If you've got a well-oiled sourcing operation running, there's every chance you'll be in a position to secure more than you can handle personally. Deal packaging is also a great activity to focus on if you are new to the game and looking to build up a cash pot to embark on your own developments as you become more experienced. By sticking to the appraisal process and securing deals under assignable contracts, you'll be in a position to successfully sell sites on to other developers and realise some cash.

The key to deal packaging is a solid appraisal, a good connection with the landowners and credible information provided to potential purchasers. Too often, we see 'deal sourcers' send over information that is lacking in detail with some of the key numbers missing, and 'exclusive' D2O deals from more than one source. To get ahead of the game, provide interested parties with a breakdown of the high-level numbers, backed up with evidence.

CASE STUDY

A client of ours was well connected with developers and investors looking for sites. With a clear idea of what his end purchasers were looking for, he pursued a number of opportunities, filtered through them using our appraisal methodology, and eventually secured a prime development site that he sold on to one of his developer contacts for a six-figure sum.

Planning uplift

Planning uplift is often incorrectly referred to as 'planning gain'. When experienced developers and consultants hear people using incorrect terminology, they immediately consider those people inexperienced.

Planning gain refers to the benefit a developer creates for the wider community by gaining planning consent on a site, for example affordable housing, improved infrastructure, and contributions for education. Planning uplift is the financial gain the developer gets as a result of obtaining planning consent on a site.

The value you create by obtaining the right planning consent is substantial, but to crystallise that profit you have to successfully sell it on to a developer, who will make their money by building out the units. It's imperative to obtain consent for a scheme that is deliverable, so when you're working out the post-planning site value, leave enough profit margin in it for the developer.

Build to sell

This is the traditional and most common exit for developers: buy or control, build or convert, then sell to make a profit. A trade strategy is market focused as you ideally want to bring units to the market at times that will maximise their value.

Off-plan sales are key to easing cash flow and creating momentum in the sales process. The emergence of technology such as computer-generated images (CGI), virtual walkthroughs, 360 cameras, time-lapse footage, drones and social media has certainly increased your ability to generate early interest and secure sales. That said, traditional marketing methods such as printed brochures, advertising boards and models are also important to securing sales.

Estate/sales agents play a big role when it comes to building to sell. Rather than negotiating them down on their fee, look to incentivise them as much as possible by increasing their earning potential in exchange for an A-rated service.

Build to rent

Building, refinancing and retaining units to rent out long term is becoming more and more popular within the development world. It gives developers something they've never really had before: recurring monthly revenue. Adding this string to your bow greatly assists with risk management of your projects. As long as you're involved in schemes that will work as rentals, the pressure to sell all units decreases.

If retaining units is your chosen exit prior to the commencement of works, it can boost your opportunity for sales (if you are planning to hold some and trade some) as your marketing can reflect the fact that some of the units are 'already reserved/sold'.

The build-to-rent model is great for long-term wealth creation, and if you are retaining units with a moderately low loan to value (LTV), you can create a good cash-flowing portfolio of properties from a single deal.

The full spectrum of exit routes gives you options for short-, medium- and long-term income generation.

🚩 RED FLAG ─────────────────────

Identify – putting all your eggs in one exit basket.

Understand – although concentrating on one specific exit strategy gives you laser focus and clarity, it can leave you at risk of being too over committed and exposed to market changes such as prices falling and changes to legislation. It's the same if you are concentrating on specific unit types.

Mitigate – the best way to beat overexposure is to diversify. With many more developers (big and small) entering the build-to-hold market as well as the more traditional build-to-sell, you can add great flexibility and allow your business to move with the market.

CHAPTER NINE

Conclusion

Property development is a 'get rich slow, get poor quick' business. In order to build a truly successful, sustainable housebuilding business, you need to start from solid foundations. The systems, strategies, insights and methods within this pocket guide are designed to focus your resources on the areas that enable you to build a low-risk, high-reward development business. Remember, look after the risks and the rewards will look after themselves.

The three main inhibitors for the majority of early stage developers are lack of clarity, no consistency and poor communication. To give your business the boost that it needs:

- Bring clarity and understanding to your purpose for being in business and the processes within the Housebuilding Business Blueprint™

- Ensure you are taking consistent action over sustained periods of time to bring rhythm to your business operations

- Communicate clearly with your internal team, external consultants and the key players in the process: land-owners, funders and the LPA.

Building a better future, together

Property development is a fantastic industry to be involved in, provided you get it right. We want to ensure developers embarking on this journey avoid the costly mistakes we've experienced ourselves and witnessed others making over the decades and generations we've been in business. Our mission is to help create development companies focused on having a positive impact on families and communities.

We are proud supporters of the UN Global Goals. By providing consultancy and support services to others, together we can help build sustainable cities and communities (goal 11) and make the world a better place in which to live, work and play.

If you'd like to know more and want to connect with us further, then please visit www.propertydevelopersacademy.com

Richard and Brynley Little.

Useful Resources

Websites

Building For Life – Builtforlifehomes.org

Designing Buildings Wiki – Designingbuildings.co.uk

EPC Register – Epcregister.com

Estates Gazette – Propertylink.estatesgazette.com

Flood Maps – Flood-map-for-planning.service.gov.uk

Flood Risk Information – Flood-warning-information
.service.gov.uk/long-term-flood-risk/map

Google Earth – Earth.google.com

Government Website – Gov.uk

Home Property Search – Home.co.uk

LandInsight – Landinsight.io

Land Registry – Landregistry-uk.com

Nimbus Maps – Nimbusmaps.co.uk

Planning Portal – Planningportal.co.uk

Planning Resource – Planningresource.co.uk

Property Development Appraisals – Pdappraisals.co.uk

Rightmove – Rightmove.com

Royal Institute of Chartered Surveyors – Rics.org

Books

Brinkley M (2017) *Housebuilder's Bible 12*. Falmouth: Red Planet Publishing

Dollard T (2017) *Designed To Perform: An Illustrated Guide To Delivering Energy Efficient Homes*. London: RIBA Publishing

Goodall M H (2017) *The Use Of Land And Buildings*. Bath: Bath Publishing

Isaac D, O'Leary J and Daley M (2010) *Property Development: Appraisal And Finance*, 2nd edition. Basingstoke: Palgrave Macmillan

Mills A (2018) *Interpreting The NPPF.* Bath: Bath Publishing

Pelsmakers S (2015) *The Environmental Design Pocketbook*, 2nd edition. London: RIBA Publishing

Reed R and Sims S (2015) *Property Development*, 6th edition. London: Routledge

The Authors

Father-and-son team Richard and Brynley Little represent the second and third generations of a family property-development business. With Tony Little starting in construction in the 1950s, over the course of six decades, the Little family has survived and thrived through multiple recessions and economic cycles, building over 3,000 homes in the process. They have worked with 100s of landowners and in collaboration with some of the UK's major housebuilders.

Since 2015, Richard and Brynley have been calling on their huge amount of experience in land, planning and development to pass on their knowledge, industry insights and best practices for building a successful property development business, offering events and consultancy services through their company Property Developers Academy. They both regularly feature in industry publications and deliver presentations at property events all over the UK.

New-build property development is pretty much all this father-and-son team has ever known. Richard in particular is strongly motivated to stop people making costly mistakes in an industry that can be both rewarding and risky. Brynley's desire is to help and inspire people in what is possible with the right knowledge, resources and support. They are renowned for their straight-talking no-BS approach, giving unfiltered real-world advice on how to make a success from creating and growing a professional housebuilding business for a better future.

Contact details

🌐 www.propertydevelopersacademy.com

📘 www.facebook.com/propertydevelopersacademy

📷 @propertydevelopersacademy

Lightning Source UK Ltd.
Milton Keynes UK
UKHW020810240519
343261UK00006B/264/P